STEAM on the road

Overleaf: The traction engine at work: Ruston Proctor No 34987, 7nhp, new in 1906, thrashing at Gressenhall, Norfolk, in September 1945. Full cut-off is being used in forward gear and the governor is controlling the speed.

SCOTTISH R
COTTISH RA
LONDON MIDL
LONDON MIDLA
AND

STEAM on the road

Anthony Beaumont

LONDON

IAN ALLAN LTD

By the same author:
Traction Engine Pictures (4th imp)
Traction Engines On Parade (2nd imp)
Rally Traction Engines (2nd imp)
Traction Engine Prints
The Organs and Engines of Thursford (3rd imp)
Steam Up! Engine and Wagon pictures (2nd imp)
Fair Organs (2nd imp)
Traction Engines and Steam Vehicles in Pictures (2nd imp)
A Gallery of Old Timers (2nd imp)
Ransomes Steam Engines — An Illustrated History
Fairground Steam
Traction Engines Past and Present

First published 1979

ISBN 0 7110 0894 9

Published by Ian Allan Ltd, Shepperton, Surrey; and printed in the United Kingdom by Ian Allan Printing Ltd

Acknowledgements

John P. Mullett for the greater part of the photographs taken during the working years of traction engines etc. T. Howe, W. Briggs, B. French, J. Crawford, C. Banham, Eddison Plant Co Ltd, J. L. Thomas for historical photographs. T. B. Paisley for information about Howards' engines. W. E. Bocking for photograph recording at rallies.

Copyrights

The copyright of all photographs originating since 1926 is vested either in one or another of the names appearing under the acknowledgements or in the author.

Note on describing cylinder dimensions

The bore is given first following by the stroke: in compound cylinders the high pressure bore is followed by the low pressure bore and then the stroke which is common to both cylinders.

Abbreviations

TE	traction engine
PE	ploughing engine
RL	road locomotive
SRL	showmans road locomotive
S tractor	showmans steam motor tractor
SC	single cylinder
SCC	single crank compound cylinders
DCC	double crank compound cylinders
pv	piston valve
psi	pounds per square inch
nhp	nominal horsepower (an arbitrary figure based only on the piston area)
bhp	brake horsepower (actual power output)
ihp	indicated horsepower

Contents

Introduction

The working era of agricultural and road steamers virtually ceased after 1945 but the full spate of their commercial use barely reached the 1930s. Now it is even more important to illustrate the various everyday tasks in which steam was the prime mover as in traction, ploughing and road engines, rollers and wagons. Their later active years have not yet passed from living memory but photographs are the best remaining link with the vital contribution that these machines made to the life of the United Kingdom in particular. Pictures of past methods of manual labour in connection with steam are significant social history.

Vested interests caused the sudden end of the highly developed steam wagons of the 1930s. They did nothing to impair the quality of life which suffers today from heavy vehicle noise. In 1976 the remaining example of a Sentinel S8 steam wagon, built in 1934, made the 1,885 mile journey London-John O'Groats-Land's End-London. Its capabilities and quiet progress deserved publicity and government observation.

The present-day steam preservation and rally movement is represented in the pictures by a variety of immaculate engines. These include unique and rare examples of special interest to technically minded readers.

In recent times it appears that there is an increasing number of engine 'conversions'. Aesthetic reasons may not be the sole motive for 'changing' say, a roller to a tractor. In any case it is debatable whether these alterations accord with genuine preservation!

Many engines in more-or-less working order were almost unsaleable in the 1930s and 1940s but prices at a 1976 auction of 'rally' engines were: £3,000 (roller) £6,000 (traction engine) and £28,500 (showmans locomotive). Unfortunately wishful first-time owners may be deterred by these figures.

The picture selection has been limited by practical considerations; nevertheless it attempts to show steam engines down the years which include an era of gathering historical interest. For this reason brief technical details are included in the captions.

South Wootton

Anthony Beaumont
October 1976

Right: Allchin (Northampton) 6 ton overtype wagon new to Northampton Borough in 1922; No 1285, Reg No NH 4463, DCC cylinders $4\frac{1}{4}$in × 8in, chain final drive and straight axle steering. Seen here in fine condition in 1947.

NORTHAMPTON CORPORAT
BOROUGH ENGINEER'S DE
Nº 4
NH-4463

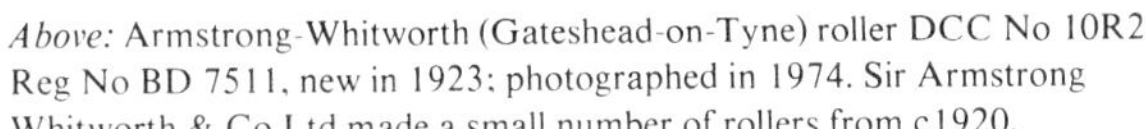

Above: Armstrong-Whitworth (Gateshead-on-Tyne) roller DCC No 10R2, Reg No BD 7511, new in 1923; photographed in 1974. Sir Armstrong Whitworth & Co Ltd made a small number of rollers from c1920.

Above right: Atkinson (Preston) chain drive tractor in Liverpool, 1947 and converted from a wagon. No Atkinsons are believed to exist in the UK. This is No 453, Reg No HS 3458, new in 1924 with duplex $6\frac{1}{4}$in × 10in cylinders, totally enclosed engine, cam-operated poppet valves and uniflow exhaust.

Right: Aveling Barford (Grantham) SC 8 ton roller No AC 606 pv, Reg No DTM 538, new in 1937 and shown here in 1973. This was the final design of steamroller to be produced in quantity. It incorporated features of Aveling Barford and Aveling & Porter.

Far right: Aveling & Porter (Rochester) vertical boiler tandem roller No 10687, Reg No XP237, with horizontal duplex cylinder engine and quick reverse valve gear. The weight could be varied from $5\frac{1}{2}$-8 ton by water ballast in the rolls. Shown at Grays, Essex in 1954.

W 425

Below: Aveling & Porter 10 ton SC roller No 3647 new to Eddisons in 1895, leaving Winchester for London as set No 20, c1900 with living van and water cart. A road scarifier is fitted to the rear axle.

Right: Aveling & Porter DCC SRL *Samson* No 4885, Reg No ME 6052; built as a RL for the Admiralty in 1901 it worked in Chatham dockyard until 1924. It was converted to a showmans by C. Presland in 1935 and so used to 1944. Shown here in 1973.

CHARLES PRESLAND. AMUSEMENT CATERER
AVELING & PORTER LTD
ROCHESTER

Right: Aveling & Porter 10 ton SC roller No 14162, new in 1933 and shown in March 1954 at Northchurch Common, Herts, when owned by Gill and Dudman, Hemel Hempstead, Herts.

Above: Aveling & Porter roller DCC pv No 12342, Reg No UT 4405, new in 1929. This was the final A&P design apart from a return to slide valves; photographed in 1973.

Right: Aveling & Porter tandem SC roller; probably a 6 ton and lightest of the range up to 15 tons. This tandem design was introduced in 1902 for quick reversing on asphalt and rolling between tramlines.

RD 4520

Right: Aveling & Porter 10 ton SC roller No 14086, Reg No VF 9891, new in 1932 to C. Banham, Runhall, Norfolk. It is shown here in 1975 before restoration and without motion covers and boiler lagging.

Below: Aveling & Porter tractor, 4nhp DCC pv No 11705, Reg No PP 7142, new in 1928. The steering gears are neatly cased in front of the boiler water tank. Photographed in 1975.

Right: Aveling & Porter DCC pv 10 ton roller No 11793, Reg No YT 1265, new in 1927 and the penultimate design. Scrapers are fitted to the rear rolls, the steering gears are uncased and the whistle is probably a restoration addition. Shown here in 1975.

YT 1265
F.R. CARNABY,
SOUTH KILLINGHOLME
6

Right: A new pair of Burrell (Thetford) chain-drive ploughing engines seen at a show in 1868 — the date of introduction. The elegant and ornamental valve chest cover will be noted besides the 'stovepipe' hat on the right.

Above: An Aveling & Porter SC roller with scarifier, shown against an interesting background in Ashley Road, Parkestone, Dorset in 1907 when working for the Eddison Steam Rolling Co.

Right: Small portable engine by Brown & May (Devizes) dating from the 1870s. It was recently discovered in a derelict and incomplete state in Wales. Originally a Watt type governor would have been fitted and not a Pickering design.

Left: Burrell DCC 6 ton side tipper wagon No 4054 or 4082 (two were supplied new to these owners in 1927/8). It was one of the last 14 Burrell wagons with $4\frac{1}{8}$in and 7in × $7\frac{1}{2}$in cylinders and 220psi boiler pressure.

Below: Burrell DCC 6 ton side tipper wagon No 4027, new to the owners in 1927. This detailed offside view shows the large final gearwheel and long chain. Ackermann steering was fitted to this and the previous wagon. No Burrell wagon appears to have survived.

Right: Burrell SC TE 8nhp made c1876 and aptly now named *Century*; probably the oldest existing Burrell TE. It is fitted with Salter safety valves, heavily weighted governor and double T-rings to the front wheels. Restored after 30 years' dereliction and shown here at Weeting, Suffolk in 1975.

CF 3667
CENTURY

Left: Burrell DCC RL *Lord Roberts* 7nhp No 3057, Reg No BE 7172, new in 1908 and first numbered 3006 by the makers before delivery. Photographed in 1973.

Below: Burrell DCC SRL No 3949 8nhp, three speed, spring mounted, *Princess Mary*. This is a works' photograph when new in 1923 for W. Nichols. It was probably the last Burrell SRL to work on the fairground when owned by C. Presland in 1958.

Right: Burrell TE DCC No 3923, Reg No EW 3026, 7nhp two speed, spring mounted, new in 1927. This engine represents the nhp size favoured for thrashing with the larger drums. Here in 1975 it is coupled to an appropriate and 'easy' load.

EW 3026
THE "BURRELL"
TRACTION ENGINE

Below: Burrell SCC roller 12½ ton No 2037, Reg No BH 8413, new in 1897. Shown in the ownership of T. T. Boughton, Amersham Common, Bucks, in 1945. The front roll yoke has been repaired in two places.

Right: Burrell SCC RL 7nhp No 2646, Reg No AD 8923, three speed, new in 1904 and shown here in 1974. This example well shows the appearance of working road engines — three-quarter canopy, disc flywheel, motion covers, boiler water tank and restrained lining out.

AD8923

Left: Burrell TE SC No 1180, 7nhp, new in 1885, and shown in its natural habitat of farm thrashing at Attleborough, Norfolk in September 1945 when owned by H. Beales. The engine has a cross-arm governor, elliptical regulator chest and cast-iron chimney.

Below: Burrell SRL DCC, 6nhp, No 3530, three speed, new in 1914. It is seen in 1949 handling an awkward load at its Thetford birthplace when owned and driven by A. T. Phoenix. The nominal 5 ton crane lift was generally exceeded.

Right: Burrell S tractor DCC, 4nhp, No 3433 *Peter Pan*, Reg No AH 0108, completed 23 December 1912 as a 5 ton tractor. Converted to a S tractor in 1932 for J. Cole, Chichester, after timber hauling in the Aldershot area since new. Photographed in 1975.

Above: Burrell SRL DCC 5nhp, No 3555 *Busy Bee*, Reg No AO 6262, three speed, new in 1914 for Taylor Bros, Amusement Caterers, Workington, Cumberland, in which area the engine worked for 26 years. It was purchased by the present owner in 1954.

Right: Burrell SRL DCC, 6nhp, three speed spring-mounted No 3836 *Starlight*, Reg No BH 8020. *Starlight* was new in 1920 as a RL in which form it remained for all its commercial work of timber-hauling, fairground and thrashing.

W. M. SALMON
FUNDENHALL
AH-0108
32
BURRELL & SONS LTD
PETER PAN
PATENT TRACTOR
BURRELL
SHOWMANS
TRACTOR

Below: Burrell SRL DCC, 8nhp, three speed, spring mounted No 3444 *His Lordship*, Reg No CK 3403, new in 1913 and used by the Green family of Glasgow and Preston, Silcock Bros of Warrington in showland. Shown in 1975 when the significance of the Stars and Stripes went unrecorded.

Right: Burrell SRL DCC special scenic, 10nhp, *Ex-Mayor* three speed, on springs. No 4000, Reg No WT 8606 was new in 1925 for G. T. Tuby & Sons, amusement caterer and former mayor of Doncaster. The cylinders are $7\frac{1}{4}$in and $11\frac{3}{4}$in × 12in, rear wheels 7ft 3in over rubbers and the working weight just over 20 tons. Shown here in 1973.

WT8606
CHARLES BURRELL & SONS LTD. · ENGINEERS · THETFORD.

Below: Clayton & Shuttleworth (Lincoln) overtype wagon with tank body and equipment for emptying drains etc. This is No 4885 on a standard 5 ton chassis, probably new in 1919 or 1920. The DCC cylinders were 4in and $7\frac{1}{4}$in × 7in and the boiler pressure 200psi.

Right: Clayton & Shuttleworth DCC roller with overhead valves and Joy valve gear. No 40102, Reg No NO 3253, fitted with scarifier and new c1907. This sale of rollers and equipment of W. G. Smoothly, Rochford, Essex, in May 1946 was typical of the end of commercial steam.

CLAYTON SHUTTLEWORTH
LINCOLN
Lot 292
NO3253
XD
8048

BH 7651
CLAYTON & SHUTTLEWORTH
AUSTIN
PV 288

Left: Clayton & Shuttleworth TE with SC $8\frac{1}{2}$in × 12in, 7nhp No 48224, Reg No BH 7651, two speed, new in 1919 and seen in 1975 (C&S were taken into Marshalls Ltd of Gainsborough in 1929). This late design is massively constructed and features a flat machined boiler mounting for the cylinder.

Above: A Clayton & Shuttleworth leading a Marshall TE towing the unusual load of a large whale on the Strang-Franade road, Isle of Man in 1921. The engines were owned by R. G. Shinner, Sulby, I of M.

Left: Cooper digger (Kings Lynn) assembled if not built in 1891 at Gt Ryburgh, Norfolk to the first design of Thomas Cooper who was then managing director of the Farmers Foundry Co. There were eight forks with 40 prongs operated from the crankshaft.

Left: Foden (Sandbach) 6 ton overtype wagon No 10788, new in 1922 with DCC cylinders $4\frac{1}{4}$in and 7in × 7in and three speeds. This wagon worked for brewers and a stone quarry company before meticulous restoration by Tate & Lyle Transport Ltd.

Below: Foden wagon, 5 ton overtype, No 6368, Reg No M 8118, built in 1916 with DCC cylinders 4in and $6\frac{3}{4}$in × 7in, restored and owned by Fodens Ltd. The name *Pride of Edwin* recalls Edwin Foden founder of the firm. The driving position gives an unobscured nearside view only.

eft: Cooper 'Fstate' digger No 13 working at Klerksdorp, S Africa. This 1901/2 design corporated a rear-mounted enclosed engine and eight forks of six prongs in the digging ame which was raised and lowered hydraulically. A neat steam steering engine is mounted the smokebox.

RO 6330
Foden
TREGUNTER
EAM HAULAGE COY
LONDON S.W.10.
65

Left: Foden 6 ton C type overtype wagon No 13488, Reg No RO 6330, new in 1927 with DCC cylinders $4\frac{1}{4}$in and 7in$\times$7in and 220psi boiler pressure. Foden's final design of four wheel overtype wagon was fitted with glass cab screens.

Below: Foden TE SC, 7nhp, No 824 *Wold Ranger*, new in 1903 and seen here in October 1944. It is hauling a thrashing drum when owned by J. Shepardson & Son, Seamer, Yorks. The small extra water tank under the boiler, large diameter rear wheels with many spokes and tall governor stand are peculiar to Foden TEs.

Above: Foden's pinnacle of steam wagon design — the 'Speed 12' undertype No 13922, built in 1930/31 with a 12 ton nominal payload, duplex poppet valve engine, integral two-speed gearbox, shaft drive and 275psi boiler. Shown at New Malden, London, in 1933.

Below: Foster (Lincoln) S tractor, No 12509, Reg No BJ 5997, DCC 4nhp new in 1910. S tractors were 5-7½ tons and used for hauling and driving medium loads such as the smaller roundabouts. Photographed in 1973.

Left: A Foster SC TE, 5nhp, in a difficult situation at Revesby Bridge, Boston, Lincs, c1925.

Left: The versatility of the TE well illustrated: the winding cables of the two engines are coupled in a twofold purchase (the withdrawn wheel pins can be seen). The Foster's steering wheel has been lashed. The LH engine appears to be a Burrell, Reg No BC 7638. The RH road locomotive's make is uncertain in this view.

PAT COLLINS, AMUSEMENTS ON TOUR
THE LEADER
DH4593
LEADER
LINCOLN

Left: Foster SRL DCC, three speed, No 14446 *The Leader*, Reg No DH 4593, new in 1920, shown working at Aston Serpentine, Birmingham Onion Fair 1950. Fosters classed these scenic type SRLs as '65bhp'. The exciter mounting behind the chimney is apparently occupied by a box.

Right: Foster SRL DCC, Class LR, 7nhp, No 14502, Reg No XC 9862 *Victory*, new in 1920, generating for a Gavioli organ in 1975. The engine was used by the amusement caterer described on the canopy. The dynamo bracket is not extended rearwards for an exciter.

Below: Foster RL DCC, 7nhp and originally a showmans for Mrs Dewhurst, Southport. Shown thrashing near Wilmslow, Cheshire in 1917 or 1918 with a labour force of land girls and German POWs — hence the smart Army private i/c.

Left: Fowler (Leeds) PE, SC, 'New 14' class, made c1876 and seen here at work in 1903 when owned by Henry Tayler, Godmanchester, Cambs. The cable shows extreme right. The vertical shaft design, bevel geared from the crankshaft and spur geared to the cable drum, remained virtually unaltered to the end of PE production in 1933.

Below: The importance of steam in agriculture in the Victorian era: 10 sets of SC Fowler PEs, cultivators and living vans, a 3nhp Fowler 'Yard' engine and a Fowler TE (partly covered) at Eddison and Allen's premises, Dorchester in 1885. 'Whites' were the usual engineers' dress.

HR4261

Left: Fowler DCC PE, Class BB1, 16nhp No 15420, Reg No HR 4261, new in 1919, and seen at work in 1975. This was the most widely used type in the UK and the cylinders are 7in and 12in×12in. The maximum direct cable pull is at least four tons and therefore over eight on a double purchase when mole-draining.

Below: The RH sister engine of the preceding Fowler PE; No 15421, Reg No HR 4262, new in 1919 and shown at the opposite side of the field in 1975. This offside view shows the automatic cable coiler arm, the two-speed gears and the signalling whistle — essential when PEs may be more than 400 yards apart.

JOHN FOWLER & Co (LEEDS) Ltd
ENGINEERS, LEEDS
DO 1927
NM 1284

Left: Fowler PEs Nos 15336 Class BB1, new in 1919 (L) and 14383, Class BB, new in 1917 (R) formerly named *Prince* at demonstration cultivating in 1973. The BB1 class have minor differences from the BB, the most easily recognisable being the guide to the LP piston rod through the front cover of the cylinder.

Right: Fowler PE Class AA7, No 15364, Reg No NO 1236, with cylinders 7$\frac{1}{2}$in and 13in × 12in, 18nhp, new in 1918. This was the largest PE type normally used in the UK. Photographed in 1975.

Left: Fowler PE, BB, No 14710, new in 1917, shown with crew when dredging near Runhall, Norfolk in 1935 when owned by C. Banham, Wood Farm, Runhall. The rear wheel rims have standard cleats fitted for soft conditions. This engine does not appear to have survived.

Below: Fowler DCC roller No 10809, Reg No DW 6015, new in 1907 and shown near York in September 1944 when owned by D. Wood & Co Ltd, Yeadon, Yorks. The boiler tank was fitted for tar and the spraying pump was chain driven from the flywheel centre.

Right: Fowler DCC roller No 16134, Reg No U 9493, Type DN1, 10 ton, photographed in 1973. Here the boiler water tank is used normally and the steam water lifter is on the tank top. The 'three wise owls' and lamb in cast brass are the Leeds' Arms.

Fowler

Left: Fowler TE SC, No 3184, new in 1876, in a thrashing set accident caused by a carelessly inserted wheel pin coming out when descending White Hill, Chesham, Bucks, on 1 January 1900. No injuries were sustained but the recovery gang looks serious about the job ahead.

Below: Fowler TE SC, 8nhp, No 7788, Reg No BH 6981(?), new in 1898, thrashing in 1958 in Bucks. The governor is practically at full throttling position which indicated a momentary easing of the thrashing load. This varied as sheaves were hand-fed into the drum.

Below: Fowler SRL DCC, three speed, *Queen Alexandra* Class A5 No 9387, Reg No SR 2234, and new in March 1904 for John Studt, Cardiff. Shown near Chesham, Bucks, between 1907 and 1915. This engine was broken up in 1941. The scrollwork decoration, open frame dynamo and steel tyres are typical of the period. Flowers grow in profusion on the verges of the macadam road.

Right: Fowler RL (recently converted to a SRL) *The Iron Duke*, 7nhp DCC, No 14758, Reg No WY 480, new in 1917 as *Empire Pride* and owned in Northumberland. The four-shaft design of Fowler engines places the crankshaft centre further forward in relation to the rear axle than Burrell's three-shaft layout. Photographed in 1975.

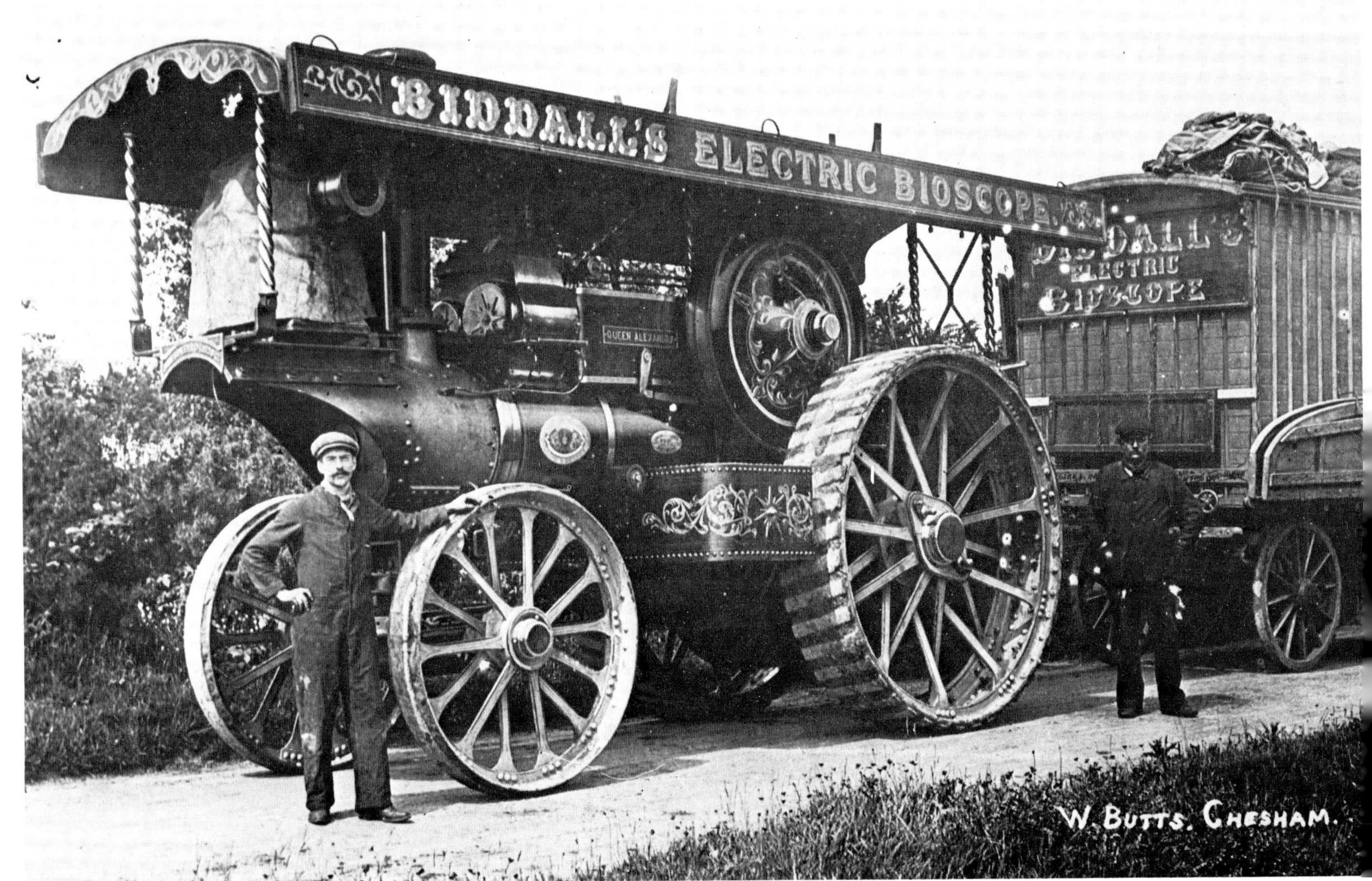

A. & G. BEALES, WISBECH, CAMBS.
THE IRON DUKE
10

Mc GIVERNS GIGANTIC PLEASURE FAIRS
DP 4418
CARRY ON

Left: Fowler SRL DCC, three-speed, Type B6, No 14425, Reg No DP 4418, *Carry On*, new in 1916 as a RL for the War Department and converted to a SRL in 1925 for Codona. Later it was used by McGivern's Amusements, N Ireland. Shown here in 1973.

Below: Supreme — the last-made Fowler SRL, a DCC B6 type with tender crane; No 20223, Reg No EU 5313, cylinders 7in and 12in × 12in rated at 10nhp with a short period maximum of 115-125ihp. The engine was new in 1934 for Mrs A. Deakin & Sons, Brynmawr, Brecon, and restored to original condition during 12 years. Photographed in 1973.

Below: Green (Leeds) SC Roller, three-shaft design, No 1399 new in 1899 shown (probably in the early 1900s) working at Harwich, Essex GER station when owned by W. G. Smoothly & Son, Rochford, Essex. The workman right, pouring a drink, has strapped trouser legs, a common habit to keep them clear of the ground.

Right: Garrett (Leiston) Agrimotor or 'Suffolk Punch', No 33180, Reg No BJ 4483, built in 1918. This DCC pv chain driven tractor of which eight were made, was virtually a final effort by steam for direct traction of agricultural implements in the field.

THE JOKER
GARRETT'S
SUFFOLK PUNCH
BJ 4483

Below: Garrett S tractor *Lord George*, No 33358, Reg No 615 CVX, DCC, 4nhp, new in 1918 as a normal Class C4 tractor for the Government and shown here in 1974. The added dynamo is smaller than normal for showland use and the canopy is well-supported by ornamental columns. Garrett's well-known crouching brass leopard is carried in front.

Right: Garrett undertype steam wagon, hydraulic tipper, No 35454, Reg No KA 9445, new in 1931, seen working for Wandsworth Gas Co, London in July 1947. This is an 8 ton capacity wagon with steam brakes, duplex cylinders $4\frac{1}{2}$in × $7\frac{1}{2}$in and cam-operated poppet valves. Apart from their six wheeler, this was Garrett's final design of wagon.

GAS
228
KR 9445

Right: Leyland (Leyland) undertype wagon No F2/72/16822, Reg No KB 8716, probably new in 1926. This final design incorporates an enclosed duplex engine with $4\frac{1}{2}$in × 6in cylinders, taper roller bearings throughout and 250psi boiler pressure. Shown here in Liverpool in May 1946.

Above: Garrett tractor DCC, 4nhp, No 33364, Reg No BH 7010, new in 1918 for the Government. Seen in 1949 when setting out from Amersham, Bucks, with a living van. The water hose is ready-coupled to the steam lifter and the damping plate for the chimney top hangs on a lamp bracket.

Right: Howard (Bedford) first four wheel 'Farmers' Engine' dated in the photograph 1873 and shown new at the works. This was almost certainly the uncompleted prototype of the cable ploughing engine with rear-mounted drum. The SC engine is in the manstand.

TATE & LYLE LTD
SUGAR REFINERS
No 2.
LIVERPOOL & LONDON
K 8716
TB 1152

BOROUGH OF
RPR

Left: Marshall (Gainsborough) 'Universal' roller No 85060, Reg No VX2427, 8/10 tons, new in 1929, photographed in Ilford, London in 1946. (Deleting nameplaces was a wartime regulation). The duplex cylinder engine and steering gear are totally enclosed and the absence of a flywheel enables quick reversing.

Above: Marshall TE SC, No 57153, new in 1911 to T.P. Saddington at Castle Donington station, Derby. Shown outside the Donington Arms. Thomas Howe was driving and it was a red-letter day with a drink for all and sundry on the peaceful road.

Right: Marshall SC TE, No 61970, Reg No BJ 5934, 7nhp two speed. This 1975 view shows the typical Marshall features of cross-arm governor, rectangular front 'spud' pan (for wheel cleats) and heavily-constructed wheels.

Below: Marshall TE SC No 72808, Reg No AH 5979, 7nhp, new in 1920 and shown driving a straw baler fed from an elevator in a 1975 demonstration. The crossed belt is not essential with a reversing engine but it provides a better grip than an open belt. The governor is maintaining a constant engine speed.

Right: Marshall SC TE No 83600, Reg No VF 3665, 7nhp, new in 1928 and photographed at Kilverstone, Thetford in 1958 when 'speeding' in full gear. It will be noticed that in the 17 years between the making dates of this engine and that on page 63, the design remained virtually unaltered.

Below: A pause in the thrashing near Diseworth, Leics in the 1980s when W. Howe was driving an early geared SC Marshall TE — probably made in the late 1870s. A young labourer kept up the water supply with buckets and the tackle was owned by T. P. Saddington.

Right: Driver Howe with a comparatively new Marshall TE — probably No 30165 made in 1898 — shown thrashing in the early 1900s at Castle Donington. The engine incorporates a bored crosshead guide which superseded parallel guide bars in some TE designs. The dogs being held up were 'ratters'.

Below: McLaren (Leeds) superheated three-speed tractor, DCC with cylinders 4½in and 7½in × 8in, photographed in France in 1915 and new in 1914. This design dates from 1910. The superheater is mounted on the smokebox and the tractor weighed just under 5 tons. A large acetylene headlamp is fitted.

Right: McLaren TE SC No 1169, Reg No JH 1941, probably new in 1912, seen driving a baler for hay near St Albans, Herts, in 1946. Again the belt is crossed for baling and the governor is in action.

Left: Ransomes Sims & Jefferies (Ipswich) SC TE, No 26991, Reg No DD 1574, new in 1916, thrashing at Redbourne, Herts, in October 1946. Two former German prisoners of war are at work. The flywheel is turning backwards and the governor is a Gardner type.

Above: Ransomes DCC TE, 7nhp, new c1910, belted to a large rack sawbench on which the saw was originally 6ft in diameter. During the pause for the photograph c1920, the safety valves have lifted. Shown near Seaford, Sussex, when owned by B. French.

Right: Robey (Lincoln) 'Express' tractor, overtype, chain drive, DCC No 43388, Reg No VL 983, new in 1929 and one of the existing two examples. Fitted with pv cylinders, roller bearing crankshaft and positive bevel/worm gear steering. Photographed in 1975.

FE 6423
MECHANICAL GROUTING CO.
47 VICTORIA ST.
ROBEY & CO LTD
LINCOLN

Far left: Robey overtype wagon seen at Bramley Depot, Surrey, in 1946. This is one of the three 6 ton 'grouting' wagons supplied new by Robeys to the Allweather Mechanical Grouting Co Ltd, Victoria, London, in 1929: DCC cylinders $4\frac{1}{2}$in and $7\frac{1}{4}$in × 7in, boiler pressure 250psi. The dangling chain earthed any 'static'.

Above: Robey tandem roller; a design introduced in the early 1920s with the same 'pistol' boiler, DCC cylinders and gearing as used in the steam wagons. The steering is direct worm geared and the chain drive is evident.

Left: Robey tractor DCC, No 37663, Reg No FE 2512, new in 1919; this is a three shaft design with cylinders $4\frac{1}{4}$in and $7\frac{1}{4}$in × 9in and 200 psi boiler pressure. Shown here at Raynes Park, Middlesex, when owned by H. Allen, Hounslow, in 1935.

AD 8889
45
ROBINSON & AUDEN LTD

Left: Robinson and Auden (Wantage) SC TE, 6nhp, No 1376, Reg No AD 8889, new in 1900; this is the only remaining TE of the small number produced by these makers between 1892 and 1900. Photographed in 1974 after recent and correct restoration.

Below: Ruston and Hornsby (Lincoln) SC roller, No 122282, Reg No BJ 9644, new in 1923 or 1924 and formerly commercially used by Alfred Dawson & Co Ltd, Rushmere, Ipswich, until 1970. This roller is fitted with two screw operated brakes — a contracting band inside the nearside wheel and a wood block on the flywheel. Photographed in 1975.

Below: Ruston 14 ton SC roller No 52848, Reg No BJ 6603, new in 1918, shown with tar sprayer at work for A. Dawson & Co Ltd in June 1948 in East Suffolk. This is a typical example of a well-kept roller even after 30 years' work.

Right: Ruthemeijer (Germany) SCC roller, No 76, Reg No 63775, new in 1907, (Charles Burrell patented a SCC design in 1889) photographed at 'Expo Steam', Peterborough in 1973 when brought over by its Dutch owner. The chimney extension is a detachable draught-improver for steam-raising.

NED. BASALT MIJ.
WL 5
FZOEF

Right: Replica of Savage's (King's Lynn) second design of front-steered SC chain TE made c1867. This one of two replicas, new in 1975 as 'Savage' continuation number 904, was made by Belmec International Ltd St Germans, King's Lynn. The second shaft is chain driven and the final double drive is by pinions and annular gears.

Left: Savage annular SCC showmans engine, probably No 614, the first of two made in 1894/5. The three in-line piston rods and their common crosshead can be seen. The outer rods were attached to the LP piston working in a cylinder which surrounded the HP cylinder. Shown new in Savage's yard with standings for fairground rides in the background.

Left: Savage TE SC, 'Sandringham' type, 8nhp, No 364, new in 1885 and broken up in 1953. Shown at Old Buckenham, Norfolk, in 1947 when named *Perfection* and owned by H. E. Loveday. Only one Savage TE is believed to exist.

Right: A technical interloper among steam vehicles but a Savage centre engine deserves an illustration. This is No 607 Type 5½, new 1894, with an organ engine which completed the motive power for roundabouts and their organs. The engine was used by Harris Bros in their gallopers.

Below: 'Super' Sentinel (Shrewsbury) undertype wagon, Type DG4P (double geared four wheels, pneumatics) No 8571, Reg No KF 6482, new in 1931. The engine has duplex cylinders 6in × 8in and the boiler is pressed to 275psi.

226

Left: Super Sentinel hydraulic tipper, No 2262, new in 1919 and shown at Wandsworth Gas Co, London, in 1947. The 'Supers' were the best known chain-drive undertypes which in the various makers' designs provided an excellent driving position compared to overtype wagons.

Below: Sentinel S4 (shaft-drive, four wheel) wagon No 9003, Reg No VE 9963, new in 1934. This is the firm's final four wheel wagon design with a four (simple) cylinder poppet valve engine mounted transversely, integral two speed gearbox and a shaft drive. Photographed in 1975.

Below: Sentinel S4 wagon No 9208, Reg No BYL 485, new in 1935 and still existing. Photographed in July 1948 at Vauxhall, London, when owned by Price & Co, Bakers. The picture shows an average load for these wagons near the end of their commercial lives and when UK steam wagon production had ceased for some 11 years.

Right: Sentinel S8T (eight wheeled tipper) No 9040, Reg No AYM 384, new in 1934 and seen here at bomb site demolition work at Nine Elms, London, in August 1942. This design was Sentinel's masterpiece having a nominal payload of over 12½ tons, shaft drive to both rear axles and steering on four wheels.

WILLMENT
BROS., LTD.
WATERLOO.
WAT. 4456.
WILLMENT
RUSTON-
BUCYRUS

Right: Super Sentinel DG4 wagon fitted for tar spraying. No 7769, Reg No MT 2805 (originally a DG6 tipper), is shown working at Raynes Park, London, in July 1947. A number of 'supers' were converted to tar sprayers.

Below: Thornycroft (Chiswick and Basingstoke) 3 ton undertype wagon, No 269, Reg No F 697, new in 1903 to Ilford Ltd and used by them to 1910. This wagon with compound cylinders 4in and 7in × 7in and boiler pressure 175psi was capable of hauling a further 2 tons.

Right: A NE Railway Co Thornycroft wagon with a top-fired locomotive type boiler. This is almost certainly an early 6 ton design of 1904, with cylinders $4\frac{1}{2}$in and 7in × 7in. A most interesting group of railway workers and others are included!

Below: Tasker (Andover) early geared three-shaft SC TE, certainly built between 1878 and 1880 and fitted with manstand steering, external hornplate pump, Watt-type governor and cast iron chimney. Shown here in Oxfordshire in 1935, apparently after long-standing dereliction when owned by D. Vanderstegan, Cane End, near Nettlebed.

Right: Tasker tractor SC No 1318, Type A1, 3nhp, 3 ton, new in 1906 and designed to take advantage of Acts which permitted one driver on vehicles under 5 ton. The cylinder is $5\frac{1}{4}$in × 9in and 22bhp was the published maximum output.

W. TASKER & SONS LTD ANDOVER ENGLAND

Right: Wallis Steevens (Basingstoke) DCC tractor, $4\frac{3}{4}$ ton fitted with oil bath motion, new in 1905 or 1906, Reg No AA 2478 and here shown at Seaham Harbour about 1910 when owned by A. French. This design was produced from 1905 to 1930.

Below: Wallis & Steevens SC 7nhp TE, new in 1924 thrashing at Milton Court Farm, Alfreston, Sussex, in September 1940 and owned by A. French, Seaford. This scene was machine-gunned by a German aircraft a few minutes after the photograph but there were no casualties.

Right: The first design of Wallis & Steevens steam wagon No 2904, 5 tons capacity, introduced in 1906 and using the same DCC engine — $4\frac{3}{4}$in and $8\frac{1}{4}$in × 9in — as fitted to the $4\frac{3}{4}$ ton tractors. There were two speeds and chain final drive. Shown new at the Basingstoke Works 1906.

S.S.M.P.H.

Right: Wallis & Steevens 'Expansion' SC TE, No 7497, Reg No NO 4710, 7nhp, new in 1916 and first used by the Ministry of Munitions. The patent expansion valve gear is operated by the governor and can be seen in this 1975 photograph.

Below: Wallis & Steevens 'Advance' roller, No 8096, Reg No ACG 760, new in 1935 and the makers' final design. The duplex cylinders and the absence of a flywheel enable instant reversing for asphalt work. Shown here in 1973.

Right: Wallis & Steevens $4\frac{3}{4}$ ton oil bath tractor operating an unusual cable ploughing attachment — probably soon after 1905. The small tender-mounted drum is spur and bevel reduction geared and the first drive is by flywheel belt.

THE "WALLIS" TRACTOR.

Left: Yorkshire (Leeds) chain-drive tractor, originally the motive section of one of the maker's six-wheeled articulated wagons. The tractor is No 2008, Reg No UA 1163, new in 1927 and shown in 1975. The double-ended boiler is here encased and the DCC cylinders are 5in and 7in × 8in.

Right: Yorkshire 2 ton capacity wagon No 117, Reg No CA 170 with undermounted DCC engine $4\frac{1}{2}$in and $7\frac{1}{2}$in × $7\frac{1}{2}$in and chain drive. This wagon is almost certainly the only remaining example of its type. The Yorkshire chain drive was introduced in 1908 and this wagon was probably new at that date.

No
CA170
STEAM CARTAGE
Wharfage & Warehousing
TEL. 322
TOM VARLEY
GISBURN STEAM CARTAGE Co
TEL 322
Pendle Maid
U.W. 2.19.0.
R.A.W. 1.10.0.
F.A.W. 1.9.0.
67
COLNE ROUND TABLE'S
GET MOVING WITH
TARTAN
A
TRANSPORT
PAGEANT
1st & 2nd SEPTEMBER
TODBER PARK

Index